RACE CAR LEGENDS

The Allisons

Mario Andretti

Crashes & Collisions

Drag Racing

Dale Earnhardt

Famous Finishes

Formula One Racing

A. J. Foyt

Jeff Gordon

The Jarretts

The Labonte Brothers

The Making of a Race Car

Monster Trucks & Tractors

Motorcycles

Richard Petty

The Pit Crew

The Unsers

Rusty Wallace

Women in Racing

CHELSEA HOUSE PUBLISHERS

RACE CAR LEGENDS

MONSTER TRUCKS & TRACTORS

Sue Mead

CHELSEA HOUSE PUBLISHERS
Philadelphia

Frontis: *Bigfoot*, the world's first monster truck, in a practice session at head-quarters in St. Louis, Missouri.

Produced by Type Shoppe II Productions, Ltd.
Chestertown, Maryland

Picture research by Joseph W. Wagner

CHELSEA HOUSE PUBLISHERS

Editor in Chief: Stephen Reginald
Managing Editor: James Gallagher
Production Manager: Pamela Loos
Art Director: Sara Davis
Photo Editor: Judy L. Hasday
Senior Production Editor: Lisa Chippendale
Publishing Coordinator: James McAvoy
Cover Illustration: Keith Trego

Cover Photos: front cover: Reuters/Ali Jarekji/Archive Photos; back cover, top: courtesy Rich Schaefer; back cover, bottom: Reuters/Archive Photos

First Printing

1 3 5 7 9 8 6 4 2

The Chelsea House Publishers World Wide Web site address is
http://www.chelseahouse.com

Library of Congress Cataloging-in-Publication Data

Mead, Sue.
    Monster trucks and tractors / Sue Mead.
        p.   cm. — (Race car legends)
    Includes bibliographical references and index.
    Summary: Looks at monster trucks and tractors, including
their history and some of the people involved in their
exhibitions and competitions.
    ISBN 0-7910-5021-1
    1. Monster trucks—History—Juvenile literature. 2. Truck
racing—Juvenile literature. 3. Farm tractors—History—Juvenile
literature. 4. Tractor driving—Competitions—Juvenile
literature. [1. Monster trucks. 2. Truck driving—Competitions.
3. Trucks. 4. Tractor driving—Competitions. 5. Tractors.] I.
Title. II. Series.
TL230.15.M43 1998
796.7—dc21
                                                    98-8216
                                                     CIP
                                                      AC

CONTENTS

1

THE MONSTERS

Bob and Dan and The Monster Trucks

Some would say that monster trucks are some of the wildest and wackiest beasts on the planet. But you probably wouldn't hear Bob Chandler say that, because he's the guy who started the whole craze.

Bob loves to tinker with cars. It all started during his teenage years when he took cars apart and put them back together again. He'd take "old junks and make 'em into street vehicles and hotrods, doing things like putting in three carburetors." He drove a pickup truck to high school and was an outcast because "it was *before* trucks were popular."

After military service and some time in college, Bob and his family took a three-month-long trip to Alaska. While driving along the rigorous Alcan Highway, Bob says, "the highway tore the tires of our 4WD camper to

Criam Forder drives his eight-ton Raging Bull *monster truck over five old cars at an American Monster Truck demonstration in Amman, Jordan.*

shreds and someone in Alaska gave me some bigger and tougher tires for the trip home."

Home was St. Louis, Missouri, where Bob had a construction job and drove a Ford F-250 4x4 for work and play. When he needed parts for his truck, there was no place close to buy them or to get repair work done, so in 1974 he opened up a 4WD shop, the Midwest Four Wheel Drive Center. For Bob, "it was the right time and the right place." He says that he never had an idea really ahead of its time, but because of his experience on the Alcan Highway, he adopted the concept of "bigger could be better," so he added bigger tires, engines, and axles to keep ahead of everybody else in the 4x4 business.

Parts at his shop sold like hotcakes as he modified his trucks and started competitions such as hill climbs. He soon realized that beating his customers wasn't smart, so he moved out of competition and into exhibition. As his trucks became "bigger and better," *Bigfoot* was born. The name actually came from Bob's shop manager, who called Bob "Bigfoot" because he couldn't keep his foot off the throttle, and sometimes broke parts on his truck. Soon, the new design wore the name.

In his first shows with *Bigfoot*, he used 48-inch tires on a Ford F-250 pickup truck. After he established a reputation with *Bigfoot*, competitors *King Kong* and *USA #1* came along, and Bob decided to go bigger again. *Bigfoot II* hit the arena with 66-inch tires, and overnight, the truck's now-famous name became a household word.

In 1981, Bob started his next adventure: car crushing. He began by driving over two cars in

a farm field for fun. Soon, this new trick caught the eye of promoters and created a craze that has skyrocketed in popularity.

Today, Bob and his fleet of 16 *Bigfoots* have been to 13 countries, but most of them are in what Bob calls Stage One of monster truck development, or in the 10,000- to 12,000-pound range. Bob has built muscle-bound 4x4s as big as 20,000 pounds. He is currently working on Stage Three trucks that will have lighter fiberglass bodies, tubular chassis, and up to 30 inches of suspension travel. The competition field has evolved to pairs, or side-by-side racing, and features trucks with 575-cubic-

Bigfoot, *painted in bright red-orange, yellow, and blue, shows its car-crushing capabilities for a crowd of cheering fans.*

inch supercharged engines. These machines produce up to 1,600 horsepower and can hit 60 miles per hour in 150 feet. Bob now has ten drivers, nine body types, video games, a web site, and a fan club with over 25,000 members. "There's very little money in racing," says Bob, "when the trucks cost between $100,000 and $150,000 each, need engine work after every race, and burn five gallons of fuel in a five-to-ten second run. "We run to win—it's the American way," he adds.

For Dan Patrick, racing tractors and trucks is just about as American as apple pie. He doesn't remember a time when the guys in his family weren't putting a horse, a tractor, or a truck on the line for a wager. "It's part of our family's history since the 1920s," he says, "a matter of who had the best horse or the best tractor."

Dan grew up on a livestock and grain farm in Kingston, Ohio, and worked with his dad 365 days a year. He quickly realized how hard the work was, and how little financial reward was in store. "When I was a child, it was natural to just pull things around the farm, and then on the weekend to go to the tractor pulls."

Dan started pulling in 1972 when he was 17, and competed at the local and state levels. In 1983, he traveled across the United States and Canada to pulling competitions and he placed seventh overall in the Modified Class. He became tired of adding and taking off weights to make the same tractor bump up in weight from 5,000 to 12,000 pounds for different events, however, and became interested in truck pulling instead.

In 1985, when tractor and truck pulls became popular and promoters asked competitors to enter, Dan was asked to build a "funny car." *Warlord* became Dan's transition into monster trucks. Over the next four years, he built two funny cars, and in 1987, built and introduced the first dragster puller.

Dan broadened his operations in 1988 and bought *Samson*, which joined the United Sports of America's Motor Spectacular as a competitor. Six months later, he was hired by Bob Chandler to build the next generation of trucks for Bob's *Bigfoot* fleet. During his

Dan Patrick's 14,000-pound 1992 monster truck, Samson, in a "monster"-ous display of power.

three years with this crew as a designer and driver, Dan was instrumental in developing and changing the industry. When he saw the need for unique trucks, each with its own personality and character, Dan decided to open his own business.

Since 1992, Dan has built 31 monster trucks. His original truck, *Warlord,* is in Orlando, Florida, at the Race Rock Cafe on permanent exhibit, and his muscle truck *Samson,* created as a 3-D design in 1993 for the television show *American Gladiators*, has performed across the country and gained popularity with each performance.

"What I do today is build the chassis for others, build component parts, and act as a consultant. In one year, I've raced forty events and built five trucks. I enjoy all aspects, but I love building the most. I want somebody to say with pride that they have a Dan Patrick truck." Involved in motor sports for over 25 years, Dan has won numerous awards, including Truck of the Year, Wreck of the Year, a Sportsmanship Award, and a Safety Award. In addition, he is the Safety and Rules Director for the Monster Truck Racing Association (MTRA).

Carl and the Monster Tractors

Carl Fowler grew up on a dairy farm in Pownal, Vermont. That meant he had to milk cows: every morning and night; every day of the week; and every week of the year. He hated to milk cows but he loved tractors, and today he lives, eats, and breathes these monster machines.

When Carl graduated from high school, he went to work for a tractor dealer, and soon he became a dealer himself. Then he began to collect tractors. He collected mostly antique, or older, models, and at one time, he had dozens of them. That meant he had to build barns to store them. "It's like a disease," he says, "once you get started, it's hard to stop."

Before long, Carl got hooked on tractor competitions, called tractor pulls. Over the years, he has pulled almost every type of tractor there is. Carl will tell you that's where the real fun begins. When he talks about competing, his 54-year-old face lights up and his eyes sparkle. He will pull out two of the trade magazines, *The Tractor Puller* and *The Hook*, and begin to tell you stories of his experiences.

Carl Fowler on his John Deere at a tractor sled pull in Sharon Spring, New York, 1995.

Carl started to compete in the Antique Class where classic models built before 1960 pull loads of up to 22,000 pounds on a stone or wooden boat. He worked his way up to the Modified Class, where tractors have as many as five engines and blowers and blast across farm fields or fan-packed arenas, yanking up to 60,000 pounds on a transfer sled. A transfer sled is a type of flat trailer hooked to the back of the tractor. As the competition begins, the weight is loaded on the rear of the sled, but as the tractor gains speed, the load is moved, or "transferred," to the front of the sled. These stylin' tractors can boast up to 1800 horsepower, reach speeds of 60 mph, and cover 300 feet in six head-snapping seconds.

No longer an active competitor, Carl now sells and repairs every-day tractors that meet a wide range of needs for their owners. What he really loves, though, is to modify tractors for the wild and wacky world of competition. If you visit him in Bennington, Vermont, you'll see a small sign along rural Route 9 that reads "Carl's Tractors." You'll notice traction machines and farm implements in the surrounding fields and on the grounds of his business. More tractors can be seen inside his shop, and upstairs, his office overflows with tractor models and tractor memorabilia. There are all kinds of vehicles in different colors, different styles, and different brands. There are posters, signs, and calendars about the rigs, and dozens of magazines and books devoted only to tractors.

Much has changed since the early days of farming when a group of local "tillers-of-the-

soil" would meet at day's end when the plowing was done to compare their horse's power
or their tractor's engine pull. "Mine's bigger or
better than yours," they would wager. Today,
there are tractor pulls at the smallest county
fairs, and at Grand National Championships
and international competitions. There are
safety regulations, panels of judges, and television coverage. Thousands sit in the driver's
seat and thousands more watch in wonder as
man and machine play out the age-old ritual
of competing against one another.

It all started with horses and traction
engines, and it is still about power and speed,
and the people who go along for the ride.

THE RACES
AND
THE RACERS

Monster truck competition and racing began with Bob Chandler and his blue Ford F-250 4x4 in St. Louis, Missouri, in 1974. In search of spare parts and fellow enthusiasts for their growing four-wheel-drive vehicle obsession, Bob and his wife Marilyn opened the Midwest Four Wheel Drive Center. They used their truck to promote their business by trying out new parts and accessories and making it bigger and better to show off to others who had 4x4s. Among other things, they added bigger and bigger tires, finally using 66-inch-tall flotation tires made by Firestone, now the standard for all monster trucks.

Named *Bigfoot*, Bob's blue pickup grew in size and fame. People loved it because it was big and unusual, and in a very short time, the Chandlers had a following of others who were building mega-machines and asking their advice. The "monster truck" phenomenon was

Bigfoot *won this Monster Truck Championship held at Madison Square Garden, New York City, on January 5, 1991.*

born. These "tricked-out" trucks competed in truck pulls, mud races, and off-road competitions that showcased their awesome power and four-wheel-drive capabilities, which allowed the trucks to do tight circles and drive almost sideways.

In 1981, Bob introduced *Bigfoot* and some friends to car crushing in a Missouri corn field. He drove his muscle-bound 4x4 over two junk cars, climbing over them and crushing them flat. It was a hit in the farm fields at home, and when he tried it a few months later at an event, the fans loved it. It wasn't long before *Bigfoot* was a star attraction at big pulls and mud-racing events, crushing cars in front of tens of thousands of screaming fans in huge stadiums.

Over the next few years, more than a hundred monster trucks entered the scene and showed up in performance arenas. When special-events promoters realized these big trucks, which looked like giant dune buggies with fiberglass bodies, were starting to gain serious attention, they scrambled to figure out how to market the growing wave of popularity. Soon, the big-tired monsters were crushing not only cars, but also limousines, buses, mobile homes, and other monster trucks. They even "popped wheelies" in addition to driving over other things. There were many contenders, like the popular *King Kong* driven by Jeff Dane, and *USA #1*, driven by Everett Jasmer, but *Bigfoot* remained the king of the monster truck world.

The next development in the monster truck competition took place in 1986 at the Houston Astrodome in an event produced by TNT Motorsports. Called monster truck racing,

trucks drove, one at a time, over an obstacle course of hills and groups of junk cars to finish with the best time. This lacked the thrill of side-by-side racing, but within two years, head-to-head competition evolved when TNT Motorsports sponsored the first national championship points series for monster trucks. Because monster truck racing was different from any other motorsport event, there were no guidelines to follow. What developed were races that awarded the winners a purse, or prize money, and competitions that became part of the point series group of races.

The basic concept for the point series, put forth in 1988, was to hold 40 races in which the muscle-bound trucks would earn points

Bear Foot, *a 13,000-pound monster truck, crushes a row of Volvos during a promotional stunt in Philadelphia.*

toward a national championship. While a few rules have changed over the past 10 years, the competition remains much the same today.

Since 1988, most monster truck competition has been in the form of drag racing. Vehicles compete in pairs in a series of elimination rounds, which leads to a final round to determine the overall champion. A typical monster truck racing event has from eight to sixteen competitors who drive their rigs in a qualifying round to achieve a qualifying time. Based on these qualifying times, drivers are paired off so that the following races have two competitors each. In each round, one truck is eliminated. The winning truck goes on to compete again until there is a final winning vehicle.

At the start of each race, drivers approach the starting line, rev up their engines, and wait for the signal to go. A set of "Christmas Tree" lights, which are red, yellow, and green, flash as signals for the start. This is called staging, or lining up the trucks for the start. While waiting for the lights, drivers torque up, or rev their engines, by putting one foot on the gas pedal and holding the truck in place with the other foot on the brake pedal. Red lighting occurs when a truck moves before the green light goes on. If the truck moves before the green light flashes on, the truck is disqualified from that day of racing. Often, the race is won or lost with the hole shot, the driver's ability to torque up and come off the starting line faster than the others.

There are straight line courses and oval courses, both with obstacles. Oval tracks are more challenging because they include turns, more obstacles, and longer driving times. Steep entrance ramps make trucks fly into the air as drivers try to clear the cars below them and land smoothly on the exit ramps. A driver's knowledge of how to keep the truck from rolling over is an important skill needed for monster truck racing. The driver has to maneuver the truck onto the entrance ramp in a way that keeps it straight in the air, then land squarely with all four tires on the exit ramp or the ground. Not taking off correctly or not landing on all four wheels can lead to a forward or sideways rollover. This occurs when a truck flips end over end or rolls over on its side. Getting big air, when the monster truck is off the ground and in the air, is a crowd-pleasing

Bigfoot *competes in a high-flying, side-by-side race in Anderson, South Carolina, in 1997.*

thrill, but if not done correctly, it can result in a rollover in which the truck breaks up or the driver is injured.

Some monster trucks have clear floors, made of lexan, to aid drivers so they can see the placement of their wheels and the surface below them. Modifications such as this and others to monster truck racing have not only changed the design of monster trucks, but also increased the need for safety standards.

Bob Chandler again led the pack in 1988 when he gathered a group of interested parties together to discuss forming a safety-oriented organization. There had been some serious injuries, and even a few deaths, in the early days of monster truck racing, and Bob wanted to prevent them from happening ever again. The Monster Truck Racing Association (MTRA) was formed to focus on safety issues for both spectators and drivers. Today, there are over a hundred monster truck owners, drivers, prompters, and sponsors that are members of the MTRA. Many safety rules have developed as a result of various mishaps. As MTRA members and officials evaluate the mishaps, they try to learn from them.

Not for men only, monster truck racing has attracted a few women as well. In 1985, Marilyn Chandler, Bob's wife and business partner, became the sport's first woman driver. She appeared on July 4th at Jack Murphy Stadium in San Diego, California, in *Ms. Bigfoot*, a Ford Ranger chassis modified to accommodate 48-inch tires and the stresses of a 571-cubic-inch, 1000-horsepower, supercharged Ford aluminum Hemi engine. It was the last *Bigfoot* vehicle to begin life as a factory production pickup truck. Two years later, *Ms. Bigfoot* was revamped with new artwork and paint, and renamed *Bigfoot Ranger*. It then made monster truck history by becoming the only Bigfoot truck ever to be sold for private, nonperformance use.

Pam Vaters of Maryland has made a name for herself as well. This well-respected professional monster truck racer is currently the only woman in the United States certified to compete in national events. To become certified, a driver must have an MTRA Class A driver's license, which is earned by performing a checklist of driving skills and having a knowledge of safety skills as another MTRA member checks and approves the applicant.

"There's a lot of work to getting a Class A license," Pam explains. "First, you get a Class B license by taking a driving test in a monster truck to show that you're capable of driving it well—doing things like accelerating, braking, and turning. Then you have to drive in ten races, sponsored by three different promotional companies, and have a Class A-licensed driver sign for each race to verify that you drove safely and competently. It's hard because you either have to have your own truck or have someone trust you to drive their truck, but it's really worth it for everyone because of the safety for the drivers and the spectators."

Pam grew up on her grandparents' farm and describes herself as a "tomboy" when she was a young girl. "I did everything the boys did, from driving tractors to motorcycles," she

Bigfoot *getting "big air."*

explains, but she never dreamed she would grow up to become a highly regarded racer or to drive the world's largest racing vehicles.

Pam married Michael Vaters, who started his career in 1982 by traveling around the country to special events and custom car shows with a built-up Ford F-250 street truck named *The Black Stallion*. The truck was set up with a 6-cylinder engine that produced 300 horsepower, and 44-inch street tires. In 1984, he started "rounding up parts from around the country" to make his tall truck a monster truck, and worked about a year and a half on the transformation. In 1986, he toured the United States with a stunt team known as the Hell Drivers and drove in exhibitions and car-crushing events.

After several injuries, Michael went to work on *The Black Stallion* to upgrade suspension components such as springs and shocks to improve suspension travel, and create a softer, gentler landing for himself and the truck. He then decided to design a monster truck using a van body, since all the other competitors at that time used pickup trucks. *Boogey Van*, equipped with an alcohol-injected, 572 cubic-inch engine, was born in Michael and Pam's garage.

This is where Pam came into the picture. "Pam used to say that she was going to be the driver of *Boogey Van* and I thought she was just kidding," Michael says. "But one day, a box showed up, and when I opened it, I pulled out a purple driving suit and saw that the blood type on it was Pam's."

Michael admits that he was nervous at first because of the risk of injury to Pam while racing, but he soon set up the cab for her. When he saw her clear three cars and land on all four wheels during her first jump, he knew Pam was good. In 1995, the pair hired a racing expert to work on the engine. It all paid off. That year, she placed fifth in the country. After putting up with teasing from some of the male drivers at the beginning of her career, she quickly began to turn heads with her winning performances. "In my first points race, I beat *Bigfoot*, which was a huge accomplishment," describes Pam, "but the best part was that Bob Chandler was happy for me. Another achievement was that I made the *Carolina Crusher* redlight one time when we were on the starting line together."

While there have been many happy moments in Pam's racing career, there have been some hard ones, too. Pam once spent four days in an intensive care unit in the hospital after running into the wall (she took out 14 feet of it) at the Pontiac Silverdome in her truck, but she was back to racing within a month. Now 32, Pam has decided to take a break and hang up her helmet in order to spend more time with son Michael Jr. on their 100-acre farm in Hagerstown, Maryland. And it's no surprise that Michael Jr. already has his eye on racing monster trucks when he gets older.

The Chandlers have stayed on the cutting edge of monster trucks and monster truck competition. There are now nearly 400 monster trucks, all of which have come to life as a result of Bob's blue Ford F-250 and the "bigger could be better" theory. This couple's dream became a true American success story and has spawned a sport now known and loved around the world.

3

TRUCKS: FACTS AND FIGURES

It takes from three months to a year to build the average monster truck. The main elements that go into changing an everyday pickup or van into a monster truck are the chassis, suspension, and horsepower. The *chassis* is the frame of a vehicle on which parts, such as the body, engine, transmission, suspension, axles, and wheels, are mounted. It has to be strong to withstand the pounding of car crushing and landing after "getting air." Today, tubular chassis with metal or steel tubes welded together are popular. They are lighter in weight and stronger than the old style frames. They are also easier to maintain and repair.

The vehicle's suspension is a mechanical system that is attached to the chassis and smoothes out the ride. It can include some or all of the following: coil springs; leaf springs; shock absorbers; airbags; and cantilevers, L-shaped hinged units that move like elbows to

Dungeon of Doom, one of the growing fleet of "Wrestletrucks," demonstrates how to drive with all four wheels off the ground.

Mark Watios of Lansing, Michigan, poses with his 13' 6", 2,000-horsepower monster truck Mongoose.

give more suspension "travel." This is particularly important because these 10,000-pound giants fly through the air and sometimes land with bone-jarring impact.

What makes these giants fly is horsepower, the unit of measure of an engine's power. All racing monster trucks can have only a single automobile engine no larger than 572 cubic inches. But mechanics can add superchargers, powerful fans that blow fuel and air into the engine to increase power. Your family car has about 125 horsepower. The first monster trucks had 500 to 1,000 horsepower. Modern monster trucks can have up to 1,600 horsepower.

The drivetrain is the system by which the wheels get power from the engine. It takes special parts in the transmission and drivetrain to use the horsepower in a safe and effective way. Most monster trucks are four-wheel drive, a system in which the engine's power goes to all four wheels. Many monster trucks also have four-wheel steering, a feature that allows the back wheels to be steered as well as the front wheels.

Other important parts include tires, body panels, and safety features. Most monster trucks use the 66-inch flotation tires made by Firestone primarily for use on big farm vehi-

A super monster truck shows its stuff to the British for the first time at the Commercial Motor Truckfest in 1986.

cles. The biggest tires used today are 10-foot-tall Alaska Tundra tires first made during WW II for land trains in Alaska. Bob Chandler has used these on two of his Bigfoot trucks. Body panels include the fenders, the cab, bedsides, hood, and tailgate. Many monster trucks use fiberglass body panels, which makes it possible to put different bodies on one chassis. The Bigfoot crew first started experimenting with body panels in 1991 and has since created many different looks and names on the tubular chassis of their fleet of trucks.

Snake Bite was the first "3D," or "character body," monster truck. Created by Bigfoot and Mattel Toys, its fiberglass body was brought to life by GTS Fiberglass of Wentzville, Missouri. *Macho Man* was inspired by Randy "Macho Man" Savage, a professional wrestler. The truck sports one of the world's largest fiberglass cowboy hats and a giant pair of sunglasses.

Because monster trucks can travel at speeds of up to 80 mph, and sometimes roll over, safety features are very important. All of the major event promoters today require that monster trucks be inspected and certified by the MTRA. This means the truck has to meet more than 100 safety requirements. All trucks have a *roll cage*, or a cage made of steel tubes that goes inside the cab to protect the driver during an accident or rollover. Since 1989, no driver using an MTRA-approved roll cage has been seriously hurt in a rollover. A Radio Ignition Interrupter (RII) is also standard on most monster trucks. This switch allows someone on the sidelines to shut off the engine.

Bob Chandler has categorized the history of monster trucks into three stages. Stage One

refers to the first generation of monster trucks, which were built for car crushing. Stage Two describes the second generation of monster trucks, which had stronger frames to survive the wilder car-crushing performances. They also had bigger engines and were easier to maneuver. Stage Three represents the third generation of monster trucks, the ones we see today. They now are being developed with a tubular chassis and a super-absorbing suspension to reduce bouncing and to provide better control and safety. They are lightweight, and have powerful engines built for racing.

Snake Bite, *with fangs ready to bite its competition, showcases its high-flying stunts at a Bigfoot race in Anderson, South Carolina.*

Nine to ten million live spectators see monster trucks each year with more than 80,000 spectators at any single event. The sport's three major promoters produce 160 to 220 events annually, representing 300 to 400 event days.

The first 4-Wheel Jamboree Nationals, considered the country's premier 4x4 event, attracted 700 participating vehicles and more than 20,000 spectators for competitions in mud bogging, truck pulling, and show-n-shine, a competition for the best-built and most attractive trucks. Sixteen years later, this event has grown to eight Jamborees a year in various locations across America.

Bigfoot (rear) buries the competition during a mud race in the movie Take This Job and Shove It.

Indianapolis, Indiana, is still the center of the 4-Wheel Jamboree Nationals and attracts more than 3,000 participating vehicles, scores of noncompeting 4x4s, and about 80,000 fans.

Bob Chandler's Bigfoot team tours America nonstop and has continuous television exposure. *Bigfoot* is the most widely seen and recognized performance vehicle in history. Children and adults alike have spent over $350 million on Bigfoot merchandise since 1982 with annual sales of over $30 million for the last eight years.

Bigfoot trucks have appeared in countless television shows and movies, including *Cannonball Run 2*, *Police Academy 6*, *Tango and Cash*, and numerous ESPN segments, TV series, and specials. The trucks have also appeared in Ford and McDonald's commercials. There are Bigfoot home videos, and the Microsoft Corporation teamed with Bigfoot in 1996 to produce a CD-ROM game with realistic, detailed simulations of monster truck driving, car crushing, and racing.

Although they no longer race for the team, some notable Bigfoot alumni impress monster truck fans. *Bigfoot 6*, built in 1986 and a record-holder for jumping 13 cars in 1987, was sold to an English promoter and still tours, under a different name, in Europe. Now an attraction at Race Rock Cafe in Orlando, Florida, the only motor sports theme restaurant in the world, *Bigfoot 7* has ten-foot-tall tires. Bigfoot trucks use many different kinds of fuel. For example, *Bigfoot 5*, which has Alaskan tundra tires, guzzles high-octane racing gas for its events. But *Bigfoot 11* and *Bigfoot 8* use methanol in a specialized "Jaz fuel cell."

An offshoot of the Bigfoot phenomenon is Wrestletrucks. The concept grew out of "monster truck sumo wrestling," an event in which two monster trucks were hooked together to try and force each other out of the arena's ring. In 1995, the new concept of Wrestletrucks, designed to create more radical action, was taken one step further, combining real wrestlers with monster trucks to create outrageous new trucks and spectacular new events.

Turner Broadcasting's World Championship Wrestling approached Bigfoot's Hollywood agents, and after several hundred hours, numerous melted truck transmissions, and thousands of dollars, a pair of trucks were de-

Wrestletrucks, such as the one shown here, are a new concept in monster truck competition, and are driven by the pro wrestlers for whom they are named.

veloped which were personalized after World Wrestling Champion's icon Hulk Hogan and the "baddest" guy on the circuit, "The Giant." The *Hulkster* and *Dungeon of Doom* were actually built on the *Bigfoot 9* and *Bigfoot 8* chassis, which were extensively remodeled, and even accommodate "copilots," or a second set of drivers. Since they race longer than regular monster trucks (about ten minutes instead of two), they are copiloted by pro wrestlers.

TRACTORS: A HISTORY

The first tractors were horse and ox teams. They were used for pulling things, but work was limited because the animals could not do more than six hours of heavy labor a day or travel more than 13 to 15 miles, and could get stuck in the fields. Spring work was typically the hardest, because the fields needed to be prepared and planted, and horses were often out of shape after the winter. In addition, horses and oxen required care and maintenance, whereas tractors only need to be fed fuel.

Nicholas Joseph Cugnot, a Frenchman, was the first person to travel in a self-propelled vehicle. In 1765, he built a machine powered by a massive steam engine that could travel six miles per hour (mph). It had three wooden wheels and a frame of heavy wooden timbers. He called it a "steam wagon," but because it was designed to pull

Mike Fitz of Melbourne, Iowa, competes in a plowing contest on August 2, 1997, with an antique tractor and plowing equipment.

Bert Rhodes drives his work horses as they pull a sickle-bar mower as Steve Pratt looks on (c. 1915).

heavy loads, it was actually the first tractor. These early developments in Europe came before similar inventions in the United States by close to a century.

Nearly 85 years later Robert Ransome of England built a "farmer's engine" designed to work in the field and operate machinery. This inventor added a steering component, which was connected to an axle, and the phenomenon was off and running.

In America, Jeremy Increase Case introduced his first steam engine in 1869. It was crude, but did the job well when compared to horses. It is no surprise, then, that the first steam tractors gained in popularity in the 1870s, and were known as "horse steering" engines. Mostly used for plowing, they were self-propelled, but not self-steering, and had "simple" and "compound" engines with one or two cylinders.

Despite their benefits, there were many difficulties. There were boiler explosions, backfiring gas, and fire hazards from the sparks released from the smokestack. It was not uncommon for a load of hay to catch fire, and there were occasional roadway accidents and injuries to people. In addition, there was the problem of startling horses not used to the commotion. This led to safety laws, one of which, in 1899, required that the operator "send a man at least 50 yards ahead to warn teams of drivers," that a steam engine was coming through.

It was Nicholas Otto's gasoline engine, built in 1876 in Germany, that started a flurry of development of the internal combustion engine in Europe and in America. Two-, three-, and four-wheeled vehicles, as well as gasoline-powered tractors, resulted. Although gas-powered engines were quickly valued for their efficiency, there were many early skeptics. By 1917, gasoline- and kerosene-powered tractors were the new state-of-the-art design, and during World War I, steam tractors were starting to go the way of ox carts and horse teams.

It was carpenter and blacksmith John Froelich, of Iowa, who developed the first gasoline-powered tractor in 1892 and a year later started the Waterloo Gasoline Traction Engine Company. He sold two tractors based on his prototype with a single-cylinder, vertical

Bert Rhodes on one of his steam tractors on his South Williamstown, Massachusetts farm (c. 1900)

engine that developed 20 horsepower. This inventor also developed reverse gearing, allowing the tractors to back up. This Waterloo Gasoline Engine Company tractor had one forward and one reverse gear. The same year, the JI Case Threshing Machine Company, of Racine, Wisconsin, entered the tractor market and developed a large, cross-motor machine. This prototype and the first Case tractor had a two-cylinder horizontal engine that produced 20 horsepower, but ignition and carburetor problems discouraged commercial promotion.

Not long after, engineering students C. W. Hart and C. H. Parr of Charles City, Iowa, were the first to use the term "tractor." In 1902, they established the Hart-Parr Company and built a 2-cylinder, 4-stroke engine that produced 30 horsepower. This tractor still ran 17 years later, an unusual feat for that time. The following year, they designed another model with water injection to prevent engine "knock." It weighed seven tons and is now on display at the Smithsonian Institution in Washington, D.C.

In 1907, Henry Ford, best known for his car and truck models, designed his first tractor, made up mostly of car parts. It sported the wheels from a binder, the 4-cylinder petrol engine from a "Model B" car, and the front axle and steering gear from a "Model K" car. This great inventor's first tractor model had a transversely-mounted engine and its four individual cylinders produced 20 horsepower. It is now on display at the Greenfield Village and Henry Ford Museum in Dearborn, Michigan. In 1915, Ford formed a new corporation,

Henry Ford and Son, to design and produce tractors on a commercial scale, and it wasn't long before he was selling them for the farm fields of America as well as exporting the well-respected Fordson tractor to Europe and Russia. These tractors were built to allow farmers to make repairs more easily and quickly.

The mergers of several companies, including McCormick and Deering, led to the birth of another well-respected tractor-maker, The International Harvester Company. Opening up shop in Chicago, Illinois, in 1902, International Harvester made one of the first tractors designed with a one-piece, cast-metal frame, and introduced innovative designs for stronger engines after the First World War. In 1924, they began selling the Farmall, which revolutionized the industry at that time.

By 1918, there were more than 140 companies selling tractors and refining designs to make more efficient and less expensive versions. It is no surprise that many inventors of this period shared concepts and designs with both motor car and tractor developers. In 1925, there were more than 500,000 tractors performing the work of over a million and a half horses and thousands of men. During this boom, many companies began to manufacture tractors, but few of them remain today.

The first tractor wheels were steel with lugs added later for better traction. Different sized lugs were bolted on for different terrain and conditions. It was a time-consuming operation, however, to attach and detach different lugs or special rims to mud-caked wheels that sometimes worked in farm fields, and at other times traveled on dirt roads. In the 1920s and

1930s, the Allis-Chalmers Company gained fame for fitting pneumatic, or rubber, tires on a rig. The Model U not only became a trend-setter, but also a record-setter, with its low-pressure, inflatable tractor tires.

A sales campaign for this tractor's speed advantage resulted in a series of publicity stunts. Barney Oldfield, who had established a reputation as an internationally recognized race car driver, drove the tractor over a mile-long course in Dallas, Texas, at an average speed of 64.28 mph. The event was officially observed by the American Automobile Association and was documented as a world speed record for agricultural tractors. Popularity spread when Allis-Chalmers hired a team of well-known racing drivers to participate in events at state fairs where modified tractors

A 1913, C. L. Best–built, 75-horsepower Tracklayer.

with the pneumatic tires, made by Firestone, raced at speeds of 30 mph and higher.

Another stunt was set up to attract the attention of farmers who attended the International Livestock Exposition in Chicago, Illinois, in 1933. A Model U was driven from Milwaukee, Wisconsin, to Chicago on the highway, a distance of 88 miles, in a head-spinning five hours and one minute. It was said that some Allis-Chalmers dealers, seeking attention, purposely drove their tractors on public roads at speeds calculated to get convictions for traveling too fast. In the end, Allis-Chalmers and low-pressure inflatable tires were a success. In 1934, the Model U was replaced by the WC, designed specifically for the new tires and priced at $825. A steel-wheeled version sold for $150 less.

During World War II, rubber was rationed, and many tractors again rode on steel wheels. But by the late 1940s, the technology that increased speed and efficiency ruled the land, and soon all tires were made from rubber.

It is interesting to note that during World Wars I and II, some tractors were cloaked with sheet metal and put into service as tanks and multipurpose vehicles. The Killen-Strait military prototype, built in 1915, is considered the first track-laying armored vehicle. Made in America, the tractor was sent to Great Britain, where it was produced as a tank and fitted with the body shell of a Delaunay-Belleville armored car.

The Daimler-Horse, built in 1915 by auto and engine developer Gottlieb Daimler to replace horses for moving war equipment, was a two-wheeled traction unit with a 4-cylinder engine that produced 14.5 horsepower.

Following the war, it was revamped for farm work and led to the Pugh motor plow, which eventually spawned the Austrian Steyr-Daimler-Puch tractor range.

Because tractor invention paralleled and interfaced with automobile development, many name plates on tractors were familiar in the car world. The "Big Three" auto makers in the U.S. today—General Motors, Ford, and Chrysler—all produced tractors, engines, or tractor parts. Ford was the most influential by far, along with inventors and carmakers like Mercedes-Benz, Fiat, and Renault.

The next major developments in ideas and inventions following the wars added comfort and convenience features. At first they were options, but they soon became standard. During the 1940s and 1950s, tractor makers such as the Minneapolis-Moline company began building tractors with cabs to protect farmers from the weather, and offered models with air conditioning, heating, and a radio. Cabbed models like the Minneapolis-Moline U-DLX brought tractors up to the level of cars by offering a fold-up passenger-side seat and gearing that allowed everyday travel of up to 40 mph. The fifth gear was a road gear that made it possible for a farmer and his family to drive into town on a Saturday night or attend church on Sunday morning. Sheet metal, along with a stylized grille, bumper, headlamps, and a cab with flip-out front and side windows, made the tractor look like a road car.

The John Deere Company, started by blacksmith and engine-plow developer John Deere at the turn of the century, and carried forward by family members, moved tractor develop-

An early John Deere tractor shows the two toed-in front wheels that make tight turns and steering easier over bumpy terrain.

ment light years ahead in the 1950s and 1960s with the development of special purpose-built tractors and multicylinder engined tractors that came with power brakes and power steering. Deere developed an attractive new green and yellow paint scheme and advertised "Power to meet every need . . . six power sizes . . . 30 basic models."

On August 29, 1960, thousands came to the Dallas Cotton Bowl for an exhibition of John Deere's new 4-cylinder engines. Mounted inside a new lineup of tractors labeled the 1010, 2010, 3010, and 4010 Series, Deere's new generation of power produced 36 to 84 horsepower, and over the next decade the company went on to sell over 400,000.

Although made earlier, four-wheel-drive versions of tractors became more popular, and were followed by tractors with increased horsepower and improved performance. In 1978, a typical "monster tractor" had the power of 747 horses and weighed 95,000 pounds when empty.

The tractor industry today is global. Tractors are used throughout the world, and parts for tractors are produced around the world. Over the years, since the days of early tractor development, many companies have bought out others to form larger companies and share technology. In the United States today, the major tractor manufacturers are John Deere, Case-International Harvester, Agco, and New Holland.

TRACTORS: FACTS AND FIGURES

Today, there are many different types of tractors. Some examples are the row crop tractor with wheels that adjust in and out for crops that grow in rows; the standard tractor, often used in wheat fields, with wheels that do not adjust; high crop tractors, taller versions used for cultivating and spraying higher crops; and orchard tractors that have pointed fenders that slide over the wheels so as not to damage the fruit, blossoms, or branches of orchard crops. An articulated tractor is a large tractor with parts that turn as well as the wheels. A chicken tractor is not a real tractor, but a chicken coop on wheels that is pulled around to spread fertilizer.

Tractors have both large and small wheels, and different tread designs on their tires. Some wheel-design types are single-wheel tractors which have one front wheel; a tricycle

One of the consistent winners in monster tractor competitions is "Fast Freddy Freeman," also known as the "Mean Mistreater," who builds high-powered competition engines from low-budget, stock materials.

tractor has two wheels close together; and the wide front end tractor has two wheels that are spread apart. Larger tractors can have from two or four tires on each side. Tractor wheels are smaller in the front than in back. Rear wheels provide the most traction and support and, therefore, need to have the greatest

Bill Hoefer fine tunes his 1929 Model D John Deere tractor at the ninth annual South Central Montana Antique Tractor and Machinery Club Threshing Bee east of Billings, Montana, in August of 1997.

amount of ground contact. Smaller front tires make steering easier and allow a smaller turning radius.

Over the past decade, four-wheel-drive tractors, with power delivered to both the front and rear wheels, have become very popular. Four-wheel-drive tractors can do more work so smaller tractors can do the job of bigger tractors. Most farming tractors start in the category of engines that produce 350 to 400 horsepower. The majority are farm-oriented, but some of these tractors are also used for industrial purposes. Caterpillar makes the "Challenger" series of crawler, or rubber-tracked, tractors that have become very popular, especially in farm fields in the west. A rubber track spreads out the weight of the tractor so that there is less compacting of the soil. Compact tractors, with 15 to 17 horsepower, are also becoming popular, and are used as "all-purpose tractors" by homeowners.

Back in the days of John Deere and his "Poppin' John's," tractor engines were fairly small, and had only two cylinders. By the 1960s, the need for increased power and speed led to the development of three-, four-, and six-cylinder engines. There are even a few eight-cylinder engines for the toughest soil and largest farms. The first tractors were two-speed with forward gears only. Today's tractors have up to 16 or 20 forward gears, and four to six gears in reverse.

When a person buys a tractor today, he or she is primarily buying the power to perform work. But other objectives, such as fuel efficiency, sound level, dealer price, and service, are important as well. Today's tractors

Dave Moore kneels beside his Super Stock John Deere 4455 tractor in Boonsboro, Maryland. Moore has been invited to a national tractor pull competition to be held in 1998, which will test how far a driver and tractor can pull a weighted sled.

have air conditioners, heaters, radios, and even cup holders for coffee or soda cans. And some have closed-circuit TV so that drivers can see the surrounding land.

All of today's tractors run on diesel fuel. After steam tractors became obsolete, a number of other low-cost fuels such as kerosene, distillate, and gasoline were used. Some tractors in the 1950s and 1960s used propane while others ran on liquid propane gas. The advantage of diesel fuel is that it develops more of a lugging, or pulling, horsepower, and a longer powerstroke. This means that the fuel

burns more slowly and helps the engine produce torque, the engine's ability to do work.
Horsepower is one of the most important but
confusing words people encounter when buying a tractor. It is simply the measure of power
an engine produces, and the term comes from
the days when work was accomplished by the
power of a horse.

How does a tractor pull its load? Early tractors used a system of belts and pulleys added
to the tractor for attaching loads or tools.
Next came the draw bar, a simple, straight

*Ralph Wiedmeyer of West Bend,
Wisconsin, rides his father's
1950 McCormick Farmall during
a field-cutting party in
September of 1997.*

bar pulled behind the tractor to which equipment was attached. The three-point hitch, a triangular-shaped hitching unit, was developed by Harry Ferguson, and today, a power take-off (PTO) is the standard method of pulling. The PTO is a shaft off the back of the tractor that operates like a driveshaft to run farming implements. A live PTO allows one power source for the tractor and a separate power source for the implements behind it. This means that the tractor can be stopped while power still operates at the rear, such as when baling hay.

All tractors must be tested before they are sold to consumers. Information about tractors built all over the world comes from the University of Nebraska's "Nebraska Test," also called the Nebraska Tractor Test Law. All specifications, data, and comparative performance results on tractors are evaluated, and fuels are compared. Because there are over 25,000 injuries and nearly one hundred deaths each year in tractor accidents, there are increasingly strict standards with regard to safety features. In the late 1960s, rollover protection was developed and is now required on all tractors to protect drivers in the event of a rollover. In the past, when a tractor rolled over onto its top or side, there was no protection for the driver. Today, a reinforced structure protects the driver and prevents the cab from collapsing.

The collecting of antique tractors has skyrocketed in popularity, and there are now collectors all over the world. Many people still run tractors that are 50 years old or more, and many of these vehicles can still do a full day's work.

While farmers have always compared tractors to see whose was the best at pulling, it wasn't until the 1950s that tractor pulls grew from a pastime into a sport. Today, there are tractor pulls all over the United States and the world. These competitions, where tractors pull different amounts of weight, have different classes and include: antique, or pre-1960 tractors; out of the field, or modern stock farm

Pat Freels, "Dollar Devil," at the wheel of his tractor with supercharged engines, spits dust into the air at the start of a transfer sled pull in front of thousands of spectators. Freels builds his monsters out of cheap throwaway parts and motors.

A six engine monster created by Tim Engler, considered "King of the Sport" in competition tractors. Engler builds almost 70 percent of the monster tractors used in competition at his shop in Princeton, Indiana.

tractors; prostock, with one turbocharger to enhance engine performance; superstock, which allows up to four turbochargers; and modified, for tractors with up to five engines and turbochargers or superchargers. Some of these engines are jet or helicopter engines.

The range of horsepower varies from 120 to 1800 with a goal of producing both torque and speed. Weights pulled range from 22,000 pounds pulled on a stone boat or wooden boat to up to 60,000 pounds pulled on a transfer sled.

Safety requirements have also increased for tractor-pulling competitions and today include seat belts, fire suits, and helmets for drivers along with roll bars, which are now becoming standard. Explosion blankets protect drivers and fans in the event that parts of the engine fly off the tractor from a malfunction or explosion, and there are shut-off switches operated by the tractor driver and the transfer sled driver in the event of an emergency.

Although tractor competition draws mostly men, there are also women who compete and race. In addition to tractor pulls, there are national and international plowing contests, and competitors often have their tractors and plows flown around the world to compete in events. Purses, or cash prizes, are offered at the highest levels of tractor competition. There are no prizes in the antique class of tractor pulling, but the superstock and modified competitions boast purses for the top points-winner (points are often gathered over the course of a year). For many, however, just as in the early days when farmers gathered in their fields to see whose horse or tractor was the best at the end of the day, the true reward is the fun of competing.

What does the future hold in store? Where it all goes from here is anyone's guess. The history of engines, the evolution of tractors, and the history of *Bigfoot* and the monster trucks have all followed a similar path, starting from the spark of invention and evolving into the spirit of challenge.

The Ford Motor Company has recently up-graded its Super Duty Truck Series for 1999. These best-selling, heavy-duty work trucks, the F250/F550 Series, have some of the world's latest and leading technology in their cabs and under their hoods. These big trucks are becoming safer with more comfort and convenience features than ever before, and are more personalized for a wide variety of needs. Henry Ford would be proud. Or would he? What do you think he would say about cupholders in a work truck, or a compartment in which to put your computer? Or what do you think he would say about monster trucks and their competitions? Perhaps he would enjoy the spectacle of engines and machinery pushed to its limits, as thousands of fans of monster trucks and tractors do today.

CHRONOLOGY

Monster Trucks

1974   Bob and Marilyn Chandler promote their new business, Midwest Four Wheel Drive Center, with their big blue Ford F-250 4x4 pickup

1978   *Bigfoot* becomes the first monster truck to use rear steering

1979   *Bigfoot* makes its first paid appearance in a car show in Denver, Colorado, and in the movie *Take This Job and Shove It*

1981   *Bigfoot 2* built, the first to use 66-inch tires; *Bigfoot*'s first car crush for an audience is a huge success

1983   The first *Bigfoot* toy is sold by Playskool, and becomes the all-time best-selling toy truck; *Bigfoot 3* built

1984   *Bigfoot 4* is built and other monster trucks hit the scene in growing numbers; *USA #1* and *King Kong* are two favorites

1985   *Bigfoot 4* stars in the largest monster truck event in history—120,000 people attend the two day event at Anaheim Stadium in California; Marilyn Chandler becomes the first female monster truck driver at the wheel of *Ms. Bigfoot*

1986   *Bigfoot 5* is built—with dual 10-foot (3-meter) Firestone tires, it is the world's tallest, widest, and heaviest pickup truck; the first event is held where monster trucks compete for time over an obstacle course, racing one at a time

1987   *Bigfoot 6* sets a long-distance car-jumping record by clearing 13 cars

1988   *Bigfoot 7* is built for the movie *Roadhouse*

1988   Side-by-side, or head-to-head, monster truck racing with both purse money and points series developed; Bob Chandler initiates the formation of the Monster Truck Racing Association, the safety and sanctioning body for this motorsport

1989   *Bigfoot 8*, the first tubular-chassis race truck, is designed, built, and tested; *Bigfoot 2* becomes the first American monster truck to tour Australia

1990   *Bigfoot 8* wins 24 of 40 races to capture monster truck racing's National Championship; *Bigfoot 3* appears in Japan at the Tokyo Dome

1991   Mattel Toys unveils its Bigfoot Champions toy line which sells out immediately

1992   Bob Chandler introduces the team concept to monster truck racing; *Bigfoot 10* wins national championship by a record margin; *Bigfoot 2* is reborn as S*afarifoot*

1993   *Bigfoot 3* is reborn as *Safarifoot 2*; *Bigfoot 11*, known as *Wildfoot*, wins the Bigfoot team's third championship

1994   Dan Patrick's *Samson* is the first monster truck to break the five-second barrier with a run of 4.983 at Bloomsburg, Pennsylvania

1995   Pam Vaters, first female in the United States certified to compete in national events, places fifth in the nation driving *Boogey Van*

1995   Wrestletrucks designed to create more radical action by combining wrestlers with monster trucks

1996   Dan Patrick awarded the first-ever Wreck of the Year Award by MTRA, resulting from the infamous rollover at Bloomsburg, Pennsylvania; Patrick also awarded the Sportsman Award

1997   Dan Patrick awarded patent from the U.S. Patent and Trademark Office in Washington, D.C., for the first 3-D body design of the *Samson* monster truck

1998   Over 10 million spectators witnessed some form of live monster truck racing action

Tractors

1765   Nicholas Joseph Cugnot of France develops first tractor, called a "steam wagon," used to transport soldiers and equipment during wartime

1849   "Farmer's engine" built by Robert Ransome in England, which had a steering component added to a crude machine design

1850s   Steam power becomes popular in United States

1869   First steam engine for farm use introduced in the United States by Jeremy Increase Case

1870s   "Horse steering" steam engines first used

1890s   Tradition and technology begin to clash, with accidents and incidents between horse teams and new steam engines on farms

1892   First successful gasoline-powered engine built by John Froelich of Iowa

1893   Waterloo Gasoline Traction Engine Company established (later bought by John Deere)

1897   Flour City one-cylinder tractor introduced

1902   Hart-Parr Company produces a two-cylinder tractor engine; still in use 17 years later

1906   International Harvester enters the scene

1918   More than 140 companies selling tractors in the United States; tractors replace 1,500,000 horses and 250,000 workers on farms

1925   More than 50,000 tractors at work on farms

1933   Pneumatic tires are introduced and become industry standard by the 1940s

1950s   Manufacturers continue to increase horsepower, include safety features, and begin to emphasize comfort by adding cabs, heaters, and air conditioners

1960s John Deere introduces multicylinder engines and tractors with power brakes and steering

1970s Increased horsepower, safety, and standard four-wheel-drive feature emphasized by all makers

1978 "Monster tractor" trend in full swing; typical size around 95,000 pounds, typical strength about 747 horsepower

1980s Tractor sales plummet because of tough times for farmers and high fuel prices

1990s Current trend is toward compact tractors for homeowners: multipurpose, with low horsepower

FURTHER READING

Johnston, Scott D. *The Original Monster Truck: Bigfoot.* Minneapolis: Capstone Press, 1994.

Leffingwell, Randy. *The American Farm Tractor: A History of the Classic Tractor.* Osceola, WI: Motorbooks International, 1991.

McKinley, Marvin. *Wheels of Farm Progress.* St. Joseph, MI: The American Society of Agricultural Engineers, 1980.

Miller, Ray H. *Fun Facts About Farm Equipment.* Dyersville, IA: The Ertl Company, Inc, 1995.

Morland, Andrew. *Thoro'bred Tractors.* London: Osprey Publishing, 1990.

Murphy, Jim. *Tractors: From Yesterday's Steam Wagons to Today's Turbocharged Giants.* New York: J.B. Lippincott, 1984.

ABOUT THE AUTHOR

Sue Wehner Mead began her automotive career as a freelance evaluator for *Four Wheeler* magazine in 1988. Today, she travels the globe test-driving cars and trucks and working as a photojournalist/feature writer for over two dozen publications. Mead specializes in 4WD and has been an off-road editor for CNN/fn. She has been a participating journalist on three Camel Trophy adventures, considered the world's most grueling off-road competition; been a codriver for the legendary off-road racer Rod Hall in the Baja 1000; and has completed three record-setting drives in stock Ford SUVs in the Arctic Circle Challenge '95, the Tip to Tip Challenge '96, and the TransAmerica Challenge '97.

ACKNOWLEDGMENTS

The author would like to thank Carl Fowler, Bob Chandler, Scott Johnston, Dan Patrick, Pam and Michael Vaters, Jim Galusha, Tara Mello, Peter MacGillivray, and Ginny Wehner. Also, a special thanks to my research assistant, Meghan Searles.

INDEX